A Walk By The Seashore

With thanks to Jennifer for all her help—C.A.
For beach-explorer Morgan—F.T.

Produced by Daniel Weiss Associates, Inc.
33 West 17 Street, New York, NY 10011

Text copyright © 1990 Daniel Weiss Associates, Inc.,
and Al Jarnow

Illustration copyright © Freya Tanz

Published by Silver Press, a division of
Silver Burdett Press, Inc., Simon & Schuster, Inc.
Prentice Hall Bldg., Englewood Cliffs, NJ 07632
For information address: Silver Press.

Printed in the United States of America
10 9 8 7 6 5 4 3 2 1

Library of Congress Cataloging-in-Publication Data
Arnold, Caroline.
A walk by the seashore/written by Caroline Arnold;
illustrated by Freya Tanz.
p. cm.—(First facts)
Summary: Walking by the seashore, a child observes
sand, waves, plants, and animals.
1. Seashore ecology—Juvenile literature.
2. Marine biology—Juvenile literature. [1. Seashore ecology.
2. Marine ecology. 3. Ecology.] I. Tanz, Freya, ill. II. Title.
III. Series: First facts (Englewood Cliffs, N.J.)
QH541.5.S35A76 1990   90-8403
574.5.2638—dc20   CIP
AC
ISBN 0-671-68666-6   ISBN 0-671-68662-3 (lib. bdg.)

A Walk By The Seashore

Written by Caroline Arnold
Illustrated by Freya Tanz

Silver Press

Feel the sand squish between your toes.
Listen to the waves splash.
Smell the salty sea air.
Where are we?
We're at the seashore.
Let's take a walk by the water.

Splash! Splash! The waves never stop moving.
Today the waves are gentle.
But sometimes they crash onto the shore.
Day after day, waves wash up rocks.
The rocks get smashed into little pieces,
until they become grains of sand.

When the wind blows grains of dry sand,
they pile up into dunes.
See the dune grasses bend in the breeze.
At the water's edge, waves shape the wet sand.
The rushing water makes tiny rivers
as it flows back to the sea.

The ocean is like a huge hole
filled with salt water.
When land rises up above the water
it forms an island.
Some islands are huge.
Others are tiny.
Look at the gull on top of the rock.
Could it be on a tiny island?

Watch as the rocks in the water disappear.
The ocean is covering them up.
That's because it is high tide.
When the ocean surface rises, it is high tide.
When the ocean surface falls, it is low tide.
There are two high tides and
two low tides each day.

You can look in your newspaper to
find out when the tide changes
where you live.

Can you see and smell the seaweed?
Seaweed is a plant that grows in the ocean.
At high tide the waves wash some on shore.
Fish and other sea animals eat seaweed.
Feel its smooth, slippery stem.

Sea Lettuce

Dead Man's Fingers

What's that jumping in the seaweed?
It's a tiny sandhopper.
Sandhoppers eat rotting seaweed.
They help to keep the beach clean.

Bladder Wrack

Irish Moss

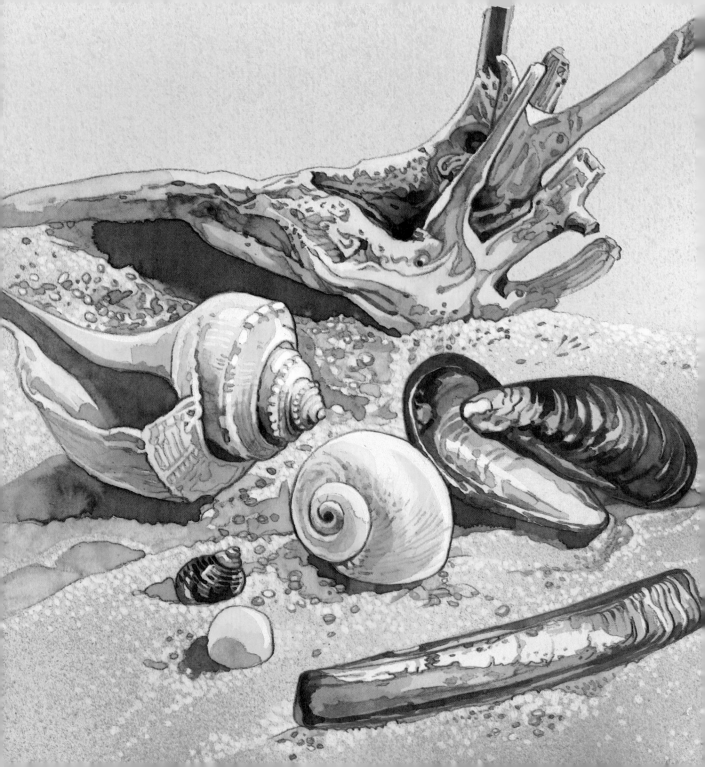

Driftwood and shells also wash up on the beach.
Pick up a shell.
Is it rough or smooth?
Clams, scallops, and mussels have two shells.
The two halves open and close.
Snails and periwinkles have only one spiral shell.

The hard shell protects the soft body of
the animal inside.
When the animal dies, its empty
shell washes up on the shore.
How many different kinds of shells can you find?

Quick! Did you see that shell move?
Underneath is a tiny hermit crab.
It lives in empty shells.
When it grows too big for a shell,
it moves to a bigger one.
Other crabs get washed up on shore, too.
Be careful that they don't pinch you!

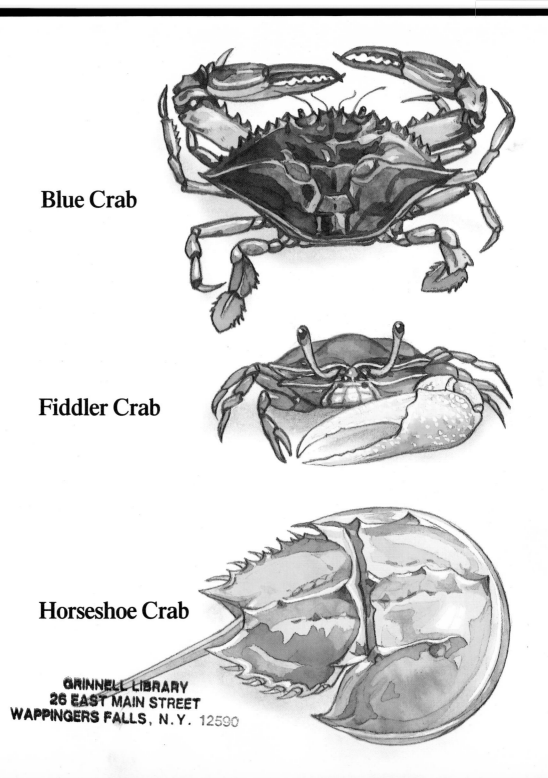

Blue Crab

Fiddler Crab

Horseshoe Crab

The tide has changed again.
Low tide leaves pools of water behind.
You can look there for crabs, living seashells,
and little fish.

Starfish live in tidepools, too.
They use their arms to move.
Turn a starfish over.
Each little bump is a tiny suction cup.
The starfish uses them to hang onto rocks,
and to pry open clams and oysters for food.
All sorts of creatures live at the sea's edge.

Skwak, *skwak*, cry the gulls.
Can you hear them as they
circle overhead?
They are looking for food in
the ocean and tidepools below.
Nearby, sandpipers are running
along the shore.
Watch them use their bills to
search the sand for tiny crabs and
worms to eat.

The birds also bring food back
to their young.
See the baby birds open their beaks
for pieces of fish.
Some birds lay their eggs right by the shore.
Others nest on rocky seaside cliffs.

Look! What is that spray of water
out in the ocean?
It's coming from a whale.
Fish get oxygen from the water.
But whales get oxygen from the air,
just like you do.

Every time a whale breathes,
it comes to the surface and
sprays water into the air.
Some whales are the biggest
animals in the world.

But in the large ocean,
even huge whales look small.
The ocean is so big that you
can't see across it.
Watch a ship sail out to sea.
Does it seem to disappear?

People once thought ships
fell off the edge of the world.
But now we know the earth is round.
The ship is just sailing across the ocean.

Many people work on the ocean.
Fishermen catch seafood for us to eat.
Deep sea divers explore
life on the ocean floor.

It's fun to play in the ocean, too.
You can swim, surf, snorkel, and windsurf.

There are seashores all over the world.

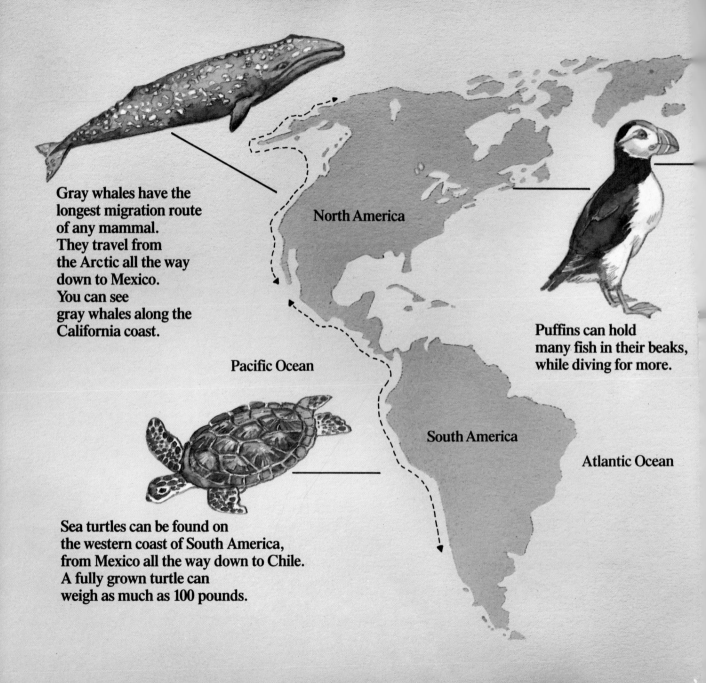

Gray whales have the longest migration route of any mammal. They travel from the Arctic all the way down to Mexico. You can see gray whales along the California coast.

North America

Pacific Ocean

Puffins can hold many fish in their beaks, while diving for more.

South America

Atlantic Ocean

Sea turtles can be found on the western coast of South America, from Mexico all the way down to Chile. A fully grown turtle can weigh as much as 100 pounds.

Do you live close to a shore?
What can you find there?

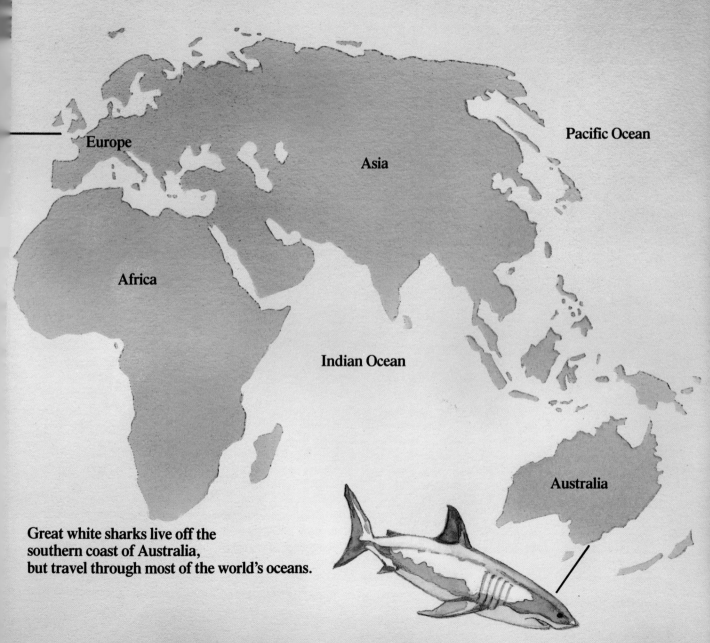

Europe

Asia

Pacific Ocean

Africa

Indian Ocean

Australia

Great white sharks live off the
southern coast of Australia,
but travel through most of the world's oceans.

What did you see on your walk by the seashore?
Here are some clues to help you remember.
Can you name each one?